Topical Conference: Green Hydrogen Production

Held at the 2024 AIChE Annual Meeting

San Diego, California, USA
27-31 October 2024

ISBN: 979-8-3313-1671-6

Printed from e-media with permission by:

Curran Associates, Inc.
57 Morehouse Lane
Red Hook, NY 12571

Some format issues inherent in the e-media version may also appear in this print version.

Printed with permission by Curran Associates, Inc. (2025)

For permission requests, please contact AIChE
at the address below.

AIChE
120 Wall Street, FL 23
New York, NY 10005-4020

Phone: (800) 242-4363
Fax: (203) 775-5177

www.aiche.org

Additional copies of this publication are available from:

Curran Associates, Inc.
57 Morehouse Lane
Red Hook, NY 12571 USA
Phone: 845-758-0400
Fax: 845-758-2633
Email: curran@proceedings.com
Web: www.proceedings.com

TABLE OF CONTENTS

GREEN HYDROGEN PRODUCTION I

GREEN HYDROGEN PRODUCTION II

358a Pathways Toward Efficient and Durable AEM Water Electrolyzers Enabled By Electro-Active Ptls

Andrew W. Tricker, Tugrul Y. Ertugrul, Jason K. Lee, Jack Lang, Iryna Zenyuk, Adam Z. Weber, Xiong Peng

Hydrogen will play a crucial role in decarbonization efforts, especially for heavy industries.[1] Durable, efficient, and low-cost electrolyzers are the core technology necessary to realize economical and green hydrogen production. Anion exchange membrane water electrolyzers (AEMWE) are an emerging technology that combines the benefits of low components cost of conventional alkaline electrolyzers and the high efficiency of proton-exchange membrane (PEM) electrolyzers.[2] A significant hurdle to the deployment of AEMWEs is durability under industrially relevant conditions. Here, we comprehensively demonstrate how porous transport layers comprised of natively active materials (PTL electrodes) can be used to achieve efficient and durable AEMWEs.[3] Specifically, stainless-steel fiber PTLs were used as anodes, without the addition of a catalyst layer. When comparing the PTL electrode to traditional catalyst layers, electrolyzer cells with several commercial anion-exchange ionomers (AEI) were subjected to current cycling. This accelerated stress testing (AST) of the electrolyzer cells showed that no significant degradation in the PTL electrode, while cell using traditional catalyst layers exhibited a loss in performance and the complete degradation of the ionomer functional group. When assessing the industrial applicability, the active area of the electrolyzer was increased 50 cm^2, which showed the same performance as the 5 cm^2 cell. The durability of the electrolyzer was tested at 2 A cm^{-2} in 0.1 M KOH for 660 h. During this period, the cell exhibited a degradation rate of 5 $\mu V\ h^{-1}$, which is comparable to typical PEM electrolyzers. The use of experiment and modeling was used to investigate the role of the structure of the PTL in influencing cell performance. The inclusion of a denser layer on the membrane side increased the activity of the electrode by decreasing the bubble coverage on the fiber surfaces and increasing the effective electrolyte conductivity within the electrode. Finally, we demonstrate a facile etching method to modify the surface of the PTL fibers to increase the active surface area of the anode and achieve a cell current density of 2.3 A cm^{-2} at 1.8 V.

References:

[1] D. S. Mallapragada, Y. Dvorkin, M. A. Modestino, D. V Esposito, W. A. Smith, B.-M. Hodge, M. P. Harold, V. M. Donnelly, A. Nuz, C. Bloomquist, *Joule* **2023**, *7*, 23–41.

[2] L. Wan, Z. Xu, Q. Xu, M. Pang, D. Lin, J. Liu, B. Wang, *Energy Environ Sci* **2023**, *16*, 1384–1430.

[3] A. W. Tricker, T. Y. Ertugrul, J. K. Lee, J. R. Shin, W. Choi, D. I. Kushner, G. Wang, J. Lang, I. V Zenyuk, A. Z. Weber, X. Peng, *Adv Energy Mater* **2024**, *14*, 2303629.

358b Understanding the Effects of Electrode Design, Cell Assembly, and Operating Conditions on AEM Electrolyzer Performance

Mahmoud Amirsalehi, Pongsarun Satjaritanun, Marta Moreno-González, Benjamin Britton, William Mustain

Hydrogen is emerging as a key energy carrier in the transition towards renewable energy, providing a pathway to achieve a zero-carbon energy landscape. While traditional technologies like the alkaline water electrolyzer (AWE) and proton exchange membrane water electrolyzer (PEMWE) dominate the commercial scene, recent developments focus on combining their benefits, leading to the creation of the anion exchange membrane water electrolyzer (AEMWE). Operating under high pH conditions, the AEMWE allows for more affordable materials and non-precious metal electrocatalysts to be used and differential pressure operation while taking advantage of an efficient zero-gap design that facilitates high-performance. Hence, the AEMWE represents a significant advancement in low-temperature water electrolysis, though further development is needed.

Anion exchange membranes, ionomers, and catalysts – as the 'active' components in AEMWE systems – have been extensively researched, with a number of materials being suggested and discussed in the literature. Yet, material selection represents only one piece of the overall puzzle when it comes to the design, operation, performance and durability of AEMWEs. Testing a zero-gap cell design in an actual electrolyzer, in contrast to half-cell or three-electrode setups, highlights how phenomena such as bubble formation, interactions among components, and mass transport shape the overall performance of the cell. Numerous factors related to electrode design, cell assembly, and operation significantly influence the performance of an electrolyzer cell.

Therefore, the primary aim of this research was to study a range of operational factors for AEMWEs and to understand how these variables influence both polarization performance and galvanostatic durability. A high-quality set of commercially available active materials including Aemion+™ anion-exchange membrane and ionomers, PtNi/C as the cathode catalyst, and either IrOx or NiFeOx as the anode catalyst, has been used in this study. Then, different electrode configurations were made by altering ionomer-to-catalyst ratios and varying the types of porous transport layers used. Furthermore, several key variables related to cell assembly were changed including the type of membrane, membrane thickness, electrode pre-treatment method, gasket thickness, and cell hardware. Finally, the operating conditions were adjusted by changing the cell break-in, temperature, heating mode, and current density.

358c Modeling Challenges in the Analysis of Proton Exchange Membrane Water Electrolysis System

Yuanxing Liu, Sahithi Srijana Akundi, Austin Braniff, Beatriz Dantas, Yuhe Tian, Shayan Niknezhad, Faisal Khan, Efstratios Pistikopoulos

Modeling Challenges in the Analysis of Proton Exchange Membrane Water Electrolysis System

Yuanxing Liu,[1,2,3], Sahithi Srijana Akundi[1,2,3], Austin Braniff[4], Beatriz Dantas[4], Shayan S. Niknezhad[1], Faisal Khan[2,3] *, Yuhe Tian[4], Efstratios N. Pistikopoulos[1,3]*

[1]*Texas A&M Energy Institute, Texas A&M University, College Station, TX, USA*

[2]*Mary Kay O'Connor Process Safety Center (MKOPSC), Texas A&M University, College Station, TX, USA*

[3]*Artie McFerrin Department of Chemical Engineering, Texas A&M University, College Station, TX, USA*

[4]*Department of Chemical and Biomedical Engineering, West Virginia University, Morgantown, WV, USA*

*Presenter E-mail: liuyuanxing@tamu.edu

Correspondence: fikhan@tamu.edu and stratos@tamu.edu

2024 AIChE Annual Meeting

October 27-31, 2024, San Diego, CA

Abstract

Proton exchange membrane water electrolysis (PEMWE) is a promising technique for the production of green hydrogen. Its exceptional performance can be attained by optimizing the geometry, materials, and operating conditions. Modeling is an effective method for predicting the performance of the cell and assessing the impacts of design and operating variables. However, there exists a discrepancy between the findings of previous studies and experimental observations in regard to the dislocation of hydrogen and oxygen evolution, as indicated by the simulation results obtained using ANSYS Fluent. This appears to be the result of the built-in electrolysis module, which mixed up the terms of exchange current densities as shown in the Scheme 1. This study presents the observation, verification, and correction of the discrepancy. Initially, the default electrolysis module was utilized to develop a CFD-based electrolysis model using ANSYS Fluent. The outcome of this model aligns with existing literature, which confirms that hydrogen is produced at the anode. A Multiphysics model of PEMWE was then developed using COMSOL to investigate the flow and electric fields. The simulation results of the COMSOL model indicate that hydrogen was produced and discharged at the cathode, which is in line with the experimental findings. The study highlights the importance of exercising caution when developing electrolysis models using the built-in module of CFD tools. Careful validation of the model against experimental data is of utmost importance to ensure its reliability and usability in practical applications.

Keywords: Proton exchange membrane water electrolysis; ANSYS Fluent; COMSOL; Simulation; Exchange current density; Hydrogen production; Hydrogen safety

For PEMFC:

$$Cathode:\ O_2 + 4H^+ + 4e^- \rightarrow 2H_2O \longrightarrow S_{O_2} = -\frac{M_{w,O_2}}{4F} R_{cat} < 0$$

$$Anode:\ H_2 \rightarrow 2H^+ + 2e^- \longrightarrow S_{H_2} = -\frac{M_{w,H_2}}{2F} R_{an} < 0$$

$$Overall:\ 2H_2 + O_2 \rightarrow 2H_2O \longrightarrow S_{H_2O} = \frac{M_{w,H_2O}}{2F} R_{cat} > 0$$

(a)

For Electrolysis:

$$Cathode:\ 2H^+ + 2e^- \rightarrow H_2 \ \overset{?}{\longrightarrow}\ S_{H_2} = \frac{M_{w,H_2}}{2F} R_{an} > 0$$

$$Anode:\ 2H_2O \rightarrow 4H^+ + O_2 + 4e^- \ \overset{?}{\longrightarrow}\ S_{O_2} = \frac{M_{w,O_2}}{4F} R_{cat} > 0$$

$$Overall:\ 2H_2O \rightarrow 2H_2 + O_2 \ \overset{?}{\longrightarrow}\ S_{H_2O} = -\frac{M_{w,H_2O}}{2F} R_{an} < 0$$

(b)

↓

For Electrolysis:

$$Cathode:\ 2H^+ + 2e^- \rightarrow H_2 \longrightarrow S_{H_2} = \frac{M_{w,H_2}}{2F} R_{an} > 0 \quad S_{H_2} = \frac{M_{w,H_2}}{2F} R_{cat} > 0$$

$$Anode:\ 2H_2O \rightarrow 4H^+ + O_2 + 4e^- \longrightarrow S_{O_2} = \frac{M_{w,O_2}}{4F} R_{cat} > 0 \quad S_{O_2} = \frac{M_{w,O_2}}{4F} R_{an} > 0$$

$$Overall:\ 2H_2O \rightarrow 2H_2 + O_2 \longrightarrow S_{H_2O} = -\frac{M_{w,H_2O}}{2F} R_{an} < 0 \quad S_{H_2O} = -\frac{M_{w,H_2O}}{2F} R_{an} < 0$$

(c)

358e Tracking Material Dynamics in Zero-Gap Water Electrolyzers Under Operando Conditions

Ryan Hannagan, Thomas Hersbach, Daniela Marin, Sang-Won Lee, Marwa Atwa, Isabela Rios Amador, Milenia Rojas Mendoza, Johanna Schroeder, Gaurav A. Kamat, Joseph Perryman, Johanna Weker, Adam Nielander, Dr. Michaela Burke Stevens, Dr. Dimosthenis Sokaras, Thomas Jaramillo

Water electrolyzers are a promising technology to electrify H_2 production. While conventional alkaline electrolysis is the most mature technology, limited capability for dynamic high current density operation necessitates a different approach. Zero-gap water electrolyzers use a solid polymer-based electrolyte membrane which enables dynamic high current density operation essential for producing high pressure and high purity H_2 from renewable energy. While promising, these electrolyzers are challenged by significant cost and durability issues that require further research and development to meet DOE cost targets of $1 per kg within the next decade.

The membrane electrode assembly (MEA) where water is converted into H_2 and O_2 is the heart of a zero-gap electrolyzer. The complex interactions between water, gas, ionomer, catalyst, and porous transport layer have a profound impact on electrochemical performance and understanding the dynamic changes to these interfaces during device operation is challenging. Here, we present a range of hardware and techniques we have developed to probe MEA dynamics under operando conditions (>2 A cm^{-2}). These include synchrotron compatible electrolyzer hardware that enables techniques such as HERFD-XANES to track the evolution of catalyst oxidation states, as well as x-ray radiography to track bubble transport. Furthermore, we have also developed lab-based methods such as online ICP-MS which enable us to track degradation resulting from material loss. Together, these methods enable insight into the dynamic chemical processes and degradation mechanisms that dictate device performance and inform design of durable next-generation water electrolyzers.

358g Production of Hydrogen By Water Splitting Using a Microwave-Assisted Reactor

Elham Ghalavand, Mustafa Erguvan, Shahriar Amini

Hydrogen has attracted considerable attention as a clean energy source due to the fact that the energy produced by the unit weight of hydrogen is considerably high. On the other hand, production of hydrogen is one of the most challenging issues in the world due to its high energy demand. Nowadays, the production of hydrogen from water is becoming very popular, since the main component of water is hydrogen which is easily accessible on the earth. A novel method to produce hydrogen is using microwave radiation due to low energy consumption [1]. During the microwave heating process, the microwave absorbent material interacts with the microwave irradiation causing a temperature increase in the sample. When microwave radiation hits the Activated Carbon the temperature rise happens, and water molecules can be broken down to its components on the high-temperature surface.

In this study, a fixed and fluidized bed quartz reactors containing activated carbon as the absorbent is utilized to dissociate water into hydrogen and oxygen under mono-mode microwave conditions. Upon reaching a specific temperature, water is introduced into the reactor from the bottom, where it decomposes upon contact with the preheated particles. Additionally, nitrogen is injected into the reactor to transport the small amount of hydrogen generated. Hydrogen levels within the system are monitored using a hydrogen sensor. The study investigates the effects of temperature variations, water flow rates, and different absorbents on hydrogen production. Results indicate that hydrogen production remains stable at 650°C, with inferior yields observed at lower temperatures compared to high temperature methods for hydrogen production.

Reference

1. Horikoshi, S., et al., *Low-temperature microwave-driven thermochemical generation of hydrogen from steam reforming of alcohols over magnetite.* International Journal of Hydrogen Energy, 2022. **47**(56): p. 23520-23529.

358h Electron Redistribution of Ru Site on MoO_2@$NiMoO_4$ Support for Efficient Ampere-Level Current Density Electrolysis of Alkaline Seawater

Yang Hou

Seawater electrolysis is a promising but challenging strategy to generate carbon-neutral hydrogen. A grand challenge for hydrogen evolution reaction (HER) from alkaline seawater electrolysis is the developing of efficient and stable electrocatalysts to overcome the limitation of sluggish kinetics. Here, we report a 3D nanorod hybrid catalyst, which comprises heterostructure MoO_2@$NiMoO_4$ supported Ru nanoparticles (Ru/ MoO_2@$NiMoO_4$) with the size of ~ 5 nm. Benefitting from the effect of strongly coupled interaction, Ru/MoO_2@$NiMoO_4$ catalyst exhibits a remarkable alkaline seawater hydrogen evolution performance, featured by a low overpotential of 184 mV at a current density of 1.0 A cm^{-2}, superior to commercial Pt/C (338 mV). Experimental observations demonstrate that the heterostructure MoO_2@$NiMoO_4$ as an electron-accepting support makes the electron transfer from the Ru nanoparticles to MoO_2, and thereby implements the electron redistribution of Ru site. Mechanistic analysis elucidates that the electron redistribution of active Ru site enhances the ability of hydrogen desorption, thereby promoting alkaline seawater HER kinetics and finally leading to a satisfactory catalysis performance at ampere-level current density of alkaline seawater electrolysis.

358i Elucidating Dynamics of Ni-Fe Based Catalyst for the Development of Zero-Gap Anion Exchange Membrane Water Electrolyzers

Milenia Rojas Mendoza, Johanna Schroeder, Ryan Hannagan, Daniela Marin, Isabela Rios Amador, Adam Nielander, Thomas Jaramillo, Dr. Michaela Burke Stevens

Zero-gap water electrolyzers utilizing anion exchange membrane (AEM) technology hold promise for efficient and sustainable hydrogen production. Due to the kinetically slow oxygen evolution reaction (OER), optimizing OER catalysts is central for optimal operation within the membrane electrode assembly (MEA). In this study, we focus on exploring varying compositions of Ni, Fe, and Co due to their promising potential as high-performance OER catalysts. Our investigation encompasses a comprehensive approach involving both fundamental catalyst characterization and device performance evaluations. Synthesizing and evaluating the performance of NiFe-based catalysts, our goal is to understand the conversion dynamics of catalyst activation to oxyhydroxide and material degradation during electrolysis. Uncovering the catalyst behavior under realistic operational conditions will provide crucial insights into the catalyst's performance under high current densities, similar to industrial conditions.

A series of metallic Ni:Fe:Co thin films were fabricated through a controlled process of electron beam physical vapor deposition and deposited onto Glassy Carbon and Porous Transport Layer substrates. This enabled precise tuning of the composition ratios to evaluate the impact on OER catalysis for a rotating disk electrode configuration and membrane electrode assembly electrolyzer. Using cyclic voltammetry in an RDE, we studied various dynamic features, such as the size and position of the Ni redox peaks. We observed that the introduction of Co inhibited the conversion of metal to oxyhydroxides, thus affecting the current density. Additionally, we found that cycling across a potential range and oxidizing and reducing the surface of the metal catalysts drove conversion, and increasing the electrolyte concentration enhanced the conversion rate of the catalyst. Furthermore, we used the same thin film deposition technique to assess the catalysts' performance in an MEA and found that while in an RDE the cobalt incorporation decreased the catalyst performance, translation to MEA configuration revealed that cobalt enhances the catalyst performance. Through this comprehensive research effort, we strive to contribute to the advancement of efficient and durable catalyst materials for next-generation AEM water electrolyzers.

Sustainable Generation of Hydrogen Using Aluminum Water Reaction Activated by Room Temperature Liquid Metal Gallium

Shyam P. Tekade[a], Diwakar Z. Shende[*b], Kailas L. Wasewar[b]

[a]Department of Chemical Engineering, Gharda Institute of Technology, Lavel, India 415708

[b]Department of Chemical Engineering, Visvesvaraya National Institute of Technology, Nagpur, India 440010

Abstract

Hydrogen is one of the important non-conventional energy sources because of its high energy content and non-polluting nature of combustions. The on-demand hydrogen gas can be produced in controllable manner through aluminum-water reaction. In the present work, the green hydrogen gas was produced on-demand using aluminum-water reaction with zero carbon emission using scrap aluminum at the ambient conditions. The experimental results are very promising and potential to make it commercialized. The experimental investigations for generation of hydrogen through aluminum-water splitting reaction have been conducted in flow conditions. The aluminum used in reaction was activated by room temperature liquid metal gallium (RTLM-Ga). The continuous hydrogen generation was attempted in flow reactor at five various flowrates of liquid reactant ranging from 0.5 to 10 mL/min. The maximum and avegarge rate of hydrogen generation were reported at the various flowrates of 0.5 N aq. NaOH in presence and absence of RTLM-Ga. The presence of RTLM-Ga was observed to enhance the rate of hydrogen generation at the tune of 1.4 – 2.6 times in comparison with that in absence of RTLM-Ga. The used RTLM-Ga was completely recovered and reused again in the activation process of scarp aluminum. The recovered RTLM-Ga was found to exhibit similar activation characteristics as that of fresh RTLM-Ga. The controlling mechanism has also been reported for the aluminum-water reaction using Shrinking Core Model.

Keywords: Hydrogen generation, aluminum-water reaction, flow reactor, gallium, Shrinking core model

Introduction

The increase in population, urbanisation, and industrialisation witnessed a peak in energy demand. The increased energy demand is fulfilled by fossil fuels, resulting in the discharge of vast amounts of various pollutants into our atmosphere during their combustion. These pollutants have many undesirable consequences on the human and environmental health. Hence, the search for new and renewable energy sources that will have minimal impact on the environment during the process of energy extraction has taken a pace in recent times. Hydrogen is one of such possible options as the clean source of energy. Hydrogen is the most abundant element on earth's surface. It is light, non-toxic, colourless, odourless and contains very high energy. Generates only water as product of combustion. Hydrogen possesses all desirable characteristics to be called as perfect fuel(Das, 1996). The characteristics like no emission of greenhouse gases, high energy content, and unlimited supply makes hydrogen the prominent candidate for future energy industry. As a fuel, hydrogen can increase the performance of automobiles, aircrafts, and utilities, keeping the environment safer and maintaining the energy security and strengthening the domestic economies of energy importing countries worldwide (Berry, et, al., 1996). A hydrogen centred energy structure is regarded as a practicable option for providing high quality energy services in a wide range of applications in an efficient, clean and safe manner while meeting sustainable objectives (Barreto, 2003). The comparison of fuel properties of hydrogen with that of gasolene and methane are listed in Table 1.

Table 1 Comparison of properties of hydrogen with other fuels (Louis and Andreas, 2001; Veziroğlu, 2003)

Property	Hydrogen	Gasoline	Methane
Density at NTP, kg/m^3	0.084	4.40	0.65
Lower Heating value, (kWh/kg)	33.33	12.4	13.9
Specific heat at constant pressure, J/gK	14.89	1.20	2.22
Ignition limits in air, vol%	4-75	1-7.6	5.3-15
Diffusion coefficient in air, cm^2/sec	0.610	0.05	0.16
Self-ignition temperature,^{0}C	585	228-501	540
Ignition energy in air, mj	0.02	0.24	0.29
Flame propagation in air, m/sec	2.65	0.4	0.4

The use of hydrogen as a fuel has various challenges related to its availability, storage and transportation. Less than 1% hydrogen is present as molecular gas. Low density, extreme flammability, wide explosive limits, low boiling point and poor compressibility make it very difficult to store or transport the hydrogen at high pressures (Tzimas, et al., 2003; Ezgi, et al., 2016; Deng, et al., 2008; Zuttel, 2003). One of the approaches to deal with the challenges of using hydrogen as a fuel is to generate hydrogen from stable and safe materials, on-demand at the location of its consumption. Development of proper method to generate hydrogen is one of the major challenges to be solved in the hydrogen economy (Lu and Chen, 2015). Hydrogen has a distinguished advantage over the other fuels as hydrogen can be manufactured from the hydrogen atom containing materials like water at or near the point of use (Berry, et., al., 1996). The major thrust in this regard is finding the materials and method with maximum capability to split the water for hydrogen generation. The reaction between aluminum and water is ideal for producing the on-demand hydrogen from stable materials. Aluminum is the most abundant metal in earth crust (Charles and William, 1990). Aluminum is cheap and environmentally safe (Huihu, et al., 2013) and commonly used in food, cosmetics and medical fields (David and Kopac 2012). But, aluminum do not react with water under atmospheric conditions because of the presence of oxide layer. The aluminum has great affinity for oxygen and results into formation of very dense oxide layer (3-9 nm) on the surface of aluminum (Yanyan, et al., 2014, Bunker, et al., 2002). Aluminum activation i.e. distortion of oxide layer is important for hydrogen generation in the aluminum-water reaction. Activation of aluminum can be possible using various methods involving physical and chemical attack on the oxide layer. The important methods explored by the various researchers include ball milling of aluminum metal, use of acids and alkali during reaction, use of oxides of various metals, use of nano-aluminum, eutectic formation of aluminum with lithium, etc. Among these, the activation of aluminum using alkali presents various advantages such as better control over the hydrogen generation rate, efficient distortion of the oxide layer, effective contact between the liquid and solid phases, cost effective, etc. The corrosive nature of the alkali underlines its use as an activator in aluminum-water reaction. It has been reported that, the room temperature liquid metal gallium has the capability for volume diffusion in the aluminum metal thus disrupting the oxide layer and hence ensures the contact between aluminum and water for generation of hydrogen. This will leads to enhancing the hydrogen generation rate during aluminum-water reaction. Moreover, the added gallium does not react with water and can be easily separated and reused in the reaction. The objectives of the present study are, to explore the capability of Gallium in hydrogen generation during aluminum-water reaction in the continuous process for hydrogen generation with the aid of gallium. Also, tuning the hydrogen generation rate with flow rate of liquid reactant and modelling the reaction using Shrinking Core Model has been attempted.

Materials and Method

The scrap aluminum wire (cylindrical, 2.4 mm diameter) has been obtained from local market. The room temperature liquid metal of gallium of AR grade (99.9% purity) was employed in the experiment. Sodium hydroxide of AR grade was used and aqueous solutions were prepared using double distilled water.

The details of experimental setup used for studying the effect of the flow rate of the liquid reactant on the aluminum-water reaction in presence of gallium have been mentioned at elsewhere (Tekade et al. 2018). Briefly, the cylindrical reactor was used for carrying the aluminum and gallium over which the liquid reactant i.e. aq. sodium hydroxide was allowed to flow at various flow rates. The experiments have been performed in absence and presence of gallium. 0.1 g of aluminum and 9.9 g of gallium were charged in the reactor. The concentration of aq. NaOH was 0.5 N throughout the experiment. The liquid reactant was injected through the reactor at five various flow rates i.e. 0.5, 1, 2, 4 and 10 mL/min to study the effect of flow rate on the hydrogen generation in aluminum-water reaction.

Results and Discussion

Effect of Flow-rate of Liquid Reactant

The flow rate of the liquid reactant was observed to have a prominent effect on the hydrogen generation rate in both cases i.e. in the absence and presence of gallium. The hydrogen generation rate was observed to increase with the flow rate of liquid reactant from 0.5 to 10 mL/min. The maximum hydrogen generation rate in absence of gallium was observed to increase from 2 mL/min to 10 mL/min whereas the average hydrogen generation rate increased from 11 mL/min to 70 mL/min with the increase in flow rate from 0.5 to 10 mL/min respectively. However, in the presence of gallium the maximum hydrogen generation rate was observed to increase from 16.9 mL/min to 21 mL/min and average hydrogen generation rate increased from 31 mL/min to 113 mL/min with the increase in flow rate from 0.5 to 10 mL/min. The data on the hydrogen generation rate at various employed flowrates is indicated in Table 2. From the data in Table 2 it has been inferred that the hydrogen generation rate can be controlled by controlling the flow rate of liquid reactant.

Table 2 Rate of hydrogen generated at various flow rates at 0.5 N NaOH and in presence of Gallium

Flow Rate, mL/min	**Hydrogen Generated, mL/min**			
	Without Ga		**With Ga**	
	Maximum	**Average**	**Maximum**	**Average**
0.5	2	11	16.9	31
1	6.6	22	11.2	32
2	6.8	36	16.9	47
4	7.5	37	26.4	98
10	10.2	70	21	113

Effect of Room Temperature Liquid Metal Gallium

The metal gallium diffuses in the aluminum and hence disturbs the oxide layer. This enhances the contact between aluminum and water and promotes hydrogen generation. This is evident from the obtained data in the current study in absence and presence of gallium. Figure 1 shows the rate of hydrogen generation with time in absence of gallium at various flowrates of liquid reactant and Figure 2 indicates the rate of hydrogen generation with time at various flowrates of liquid reactant in presence of gallium. Also, the data in Table 2 highlights the effect of presence of gallium in hydrogen generation in aluminum-water reaction. The rate of hydrogen generation was observed to increase at every employed flowrate of liquid reactant in presence of gallium as compared to that in absence of gallium. In the present study the intensification in the range of 1.4 - 2.6 times the rate of hydrogen generation is observed because of the gallium in reaction. The theoretical amount of energy generated in the absence and presence of gallium in the reaction based on average hydrogen generation rate at various employed flowrates of liquid reactant has been computed and given in Table 3.

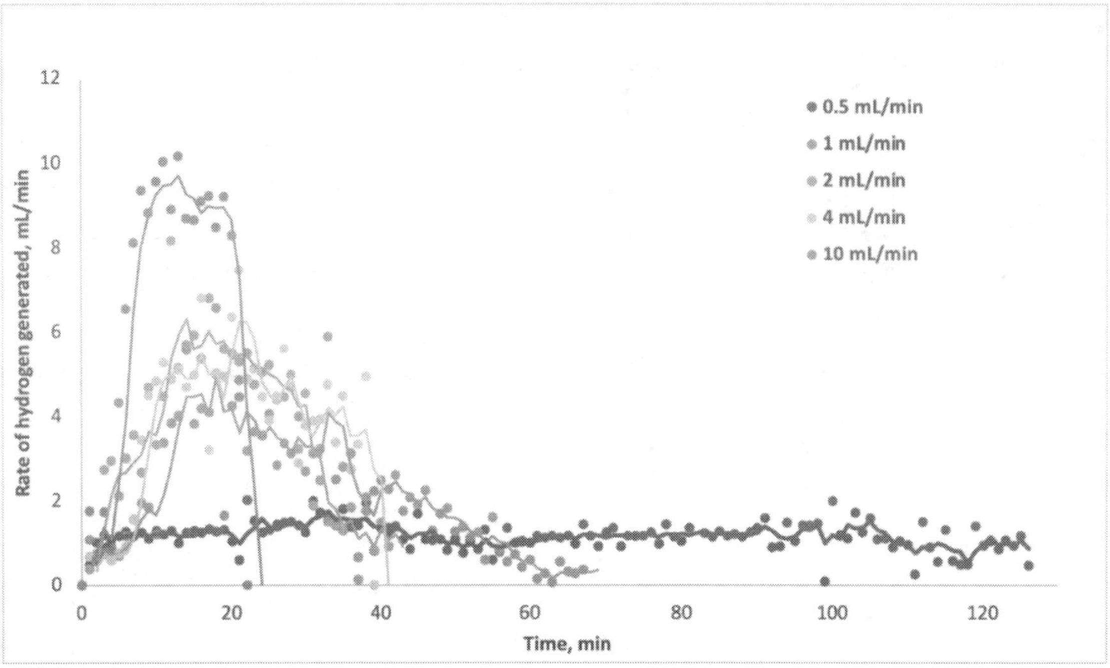

Fig. 1 Rate of hydrogen generated at various flow rates at 0.5 N NaOH and in the absence of Gallium

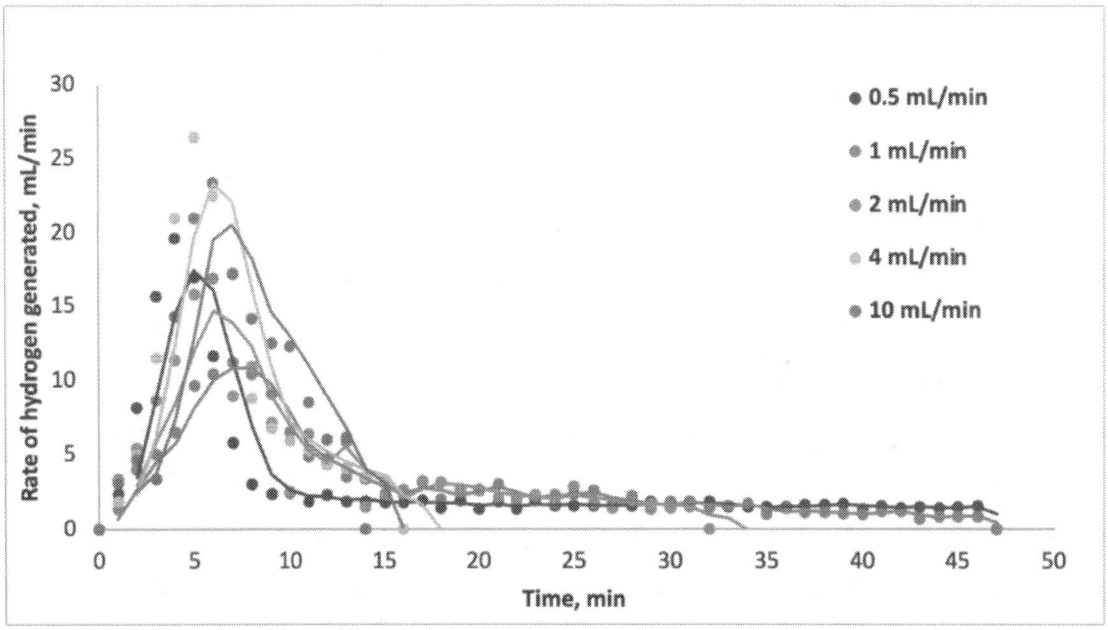

Fig. 2 Rate of hydrogen generated at various flow rates at 0.5 N NaOH and in the presence of Gallium

Table 3 Theoretical Amount of Energy Generated as Per Average Hydrogen Generation Rate

Flow Rate, mL/min	**Energy Generated, W**	
	Without Ga	**With Ga**
0.5	3.135	6.022
1	4.138	6.022
2	6.674	8.943
4	6.923	18.486
10	13.2	21.32

Shrinking Core Model

The reaction between aluminum and water is heterogenous reaction and hence can have the multiple controlling mechanism (mass transfer/chemical reaction). Shrinking core model can be used for predicting the controlling mechanism for heterogeneous reactions. Here, the predicted conversion were determined using the appropriate model equation of SCM and the predicted valued are compared with the obtained experimental conversion of the aluminum in the reaction. The following model equation with the corresponding controlling step are given below,

For Diffusion through Liquid Film Controls (Shrinking Particle),

$$X_B = 1 - \left(1 - \frac{t}{\tau}\right)^{1.5}$$

For Chemical Reaction Controls (Shrinking Particle),

$$X_B = 1 - \left(1 - \frac{t}{\tau}\right)^{2}$$

For Diffusion through Liquid Film Controls (Constant Particle Size),

$$X_B = \frac{t}{\tau}$$

Where X_B is the predicted conversion; t is the time at which the given conversion is obtained and τ is the time required for complete conversion corresponds to $X_B = 1$.

The theoretical conversion obtained from the above equation at respective time has been compared with the experimental conversion and the coefficient of determination R^2 was obtained as follows,

$$R^2 = 1 - \frac{\text{sum squared regression (SSR)}}{\text{total sum of squares (SST)}}$$

$$SSR = \Sigma(X_B - X'_B)^2$$

$$SST = \Sigma(X_B - X_{BM})^2$$

Where, X_B is actual experimental conversion; X'_B is predicted conversion; and X_{BM} is the mean of actual conversion.

The values of the coefficient of determination at employed experimental conditions are listed in Table 4. The value of R^2 close to 1 indicates the ideal experimental data fitting to the considered model. Hence, the maximum value of R^2 may be used for predicting the controlling mechanism. The predicted controlling mechanism is also listed in Table 4. The aluminum-water reaction in the absence of gallium was observed to have a chemical reaction as rate-controlling step at all flowrates of liquid reactant except 4 mL/min. The rate of diffusion of liquid reactant through the oxide layer may be fast as compared to the rate of chemical reaction between aluminum and water. Also, the flow of liquid past the aluminum surface may carry the solid byproduct and hence prevents the buildup of the solids at the aluminum surface. However, in the presence of gallium it has been observed that SCM has predicted diffusion as rate controlling mechanism for lower flowrates of 0.5, 1 and 2 mL/min whereas chemical reaction as rate controlling mechanism for 4 and 8 mL/min. The presence of gallium in the reaction may create the extra layer over the surface of aluminum. The liquid reactant has to diffuse through this gallium layer so as to reach to the surface of aluminum and therefore decreasing the rate of diffusion as compared to the rate of a chemical reaction. With the increase in flowrate of liquid reactant, more and more liquid reactant now available for the diffusion and hence increasing the rate of diffusion over the rate of chemical reaction.

Table 4 Controlling Step According to Shrinking Core Model

Flow rate, mL/min		Coefficient of Determination, R^2			Controlling Step
		Chemical Reaction controls	**Diffusion through liquid film controls (changing size)**	**Diffusion through liquid film controls (constant size)**	
0.5	Without Ga	**0.974**	0.794	0.837	Reaction
	With Ga	0.755	0.599	**0.923**	Diffusion
1	Without Ga	**0.870**	0.695	0.855	Reaction
	With Ga	0.787	0.629	**0.925**	Diffusion
2	Without Ga	**0.909**	0.74	0.822	Reaction
	With Ga	0.793	0.626	**0.92**	Diffusion
4	Without Ga	**0.926**	0.835	**0.75**	Diffusion
	With Ga	**0.850**	0.681	0.846	Reaction
10	Without Ga	**0.9**	0.863	0.717	Reaction
	With Ga	**0.918**	0.787	0.775	Reaction

Conclusions

Scrap aluminum was successfully utilized for continuous generation of hydrogen in aluminum-water reaction. Hydrogen generation rate can be tuned by tuning the flowrate of liquid reactant flowing through the reactor. Intensification in the rate of hydrogen generation was in the range of 1.4 - 2.6 times due to the Gallium. The presence of Gallium in the reaction leads to the diffusion of liquid reactant as the controlling step according to SCM.

References

1. Barretoa L., Makihiraa A. and Riahia K.,. The hydrogen economy in the 21st century: a sustainable development scenario, Int. J. Hydrogen Energy, 28, 267-284, 2003.
2. Bunker B.C, Nelson G.C, Zavadil K.R, Barbour J.C, Wall F.D, Sullivan J.P, C. F. Windisch Jr., M. H. Engelhardt, and D. R. Baer,. "Hydration of passive oxide films on aluminum", J Phys Chem B, 106, 4705- 4713, 2002.
3. Charles T. D., and William D. S, The chemistry of aluminum in the environment, Environmental Geochemistry and Health., 12, 28–49, 1990.
4. David E., and Kopac J. Hydrolysis of aluminum dross material to achieve zero hazardous waste, J. Hazardous Materials, 209–210, 501–509. 2012.
5. Das L. M., Hydrogen-oxygen reaction mechanism and its implication to hydrogen engine combustion, Int. J. Hydrogen Energy, 21, 703–715, 1996.
6. Deng Z. Y., Ferreira J. M. F., and Sakka Y., Hydrogen-generation materials for portable applications, J Am Ceram Soc., 91, 3825-34, 2008.
7. Ezgi D., Tekkaya., and Yuda Y., Mesoporous MCM-41 material for hydrogen storage: A short review, Int. J. Hydrogen Energy, 41, 9789–9795, 2016.
8. Huihu W., Ying C., Shijie D., Zhifeng L., Qingbiao Z., Ping L., and Zhixiong X., Investigation on hydrogen production using multicomponent aluminum alloys at mild conditions and its mechanism, Int. J. Hydrogen Energy., 38, 1236-1243, 2013.
9. Levenspiel Octave. Chemical reaction engineering. 3rd ed. New York: John Wiley and sons; 1999.
10. Louis S., and Andreas Z., Hydrogen storage materials for mobile applications, Nature, 414, 353-358, 2001.
11. Nicoletti G., Arcuri N., Nicoletti G., and Bruno R., A technical and environmental comparison between hydrogen and some fossil fuels, Energy Conversion Management, 89, 205–213, 2015.
12. Tekade SP, Shende DZ, Wasewar KL. Hydrogen generation in an annular micro-reactor: an experimental investigation and reaction modelling by Shrinking Core Model (SCM). Int J Chem React Eng 2018;16:7.
13. Tzimas E., Filiou C., Peteves S. D., and Veyret J. B. Hydrogen storage: state-of- the-art and future perspective, European commission document, Institute For Energy Pattern, The Netherlands. 2003.

14. Veziroğlu T. N., 21st century's energy: hydrogen energy system, Pamukkale Üniversitesi Mühendislik Fakültesi, Denizli, 2003.

15. Yanyan J., Jie S., Haixia M., Yamin D., Yujun C., and Ning W., Hydrogen generation using a ball-milled Al/Ni/NaCl mixture, J. Alloys Compounds, 588, 259–264. 2014.

16. Zuttel A., Materials for hydrogen storage, Materials Today, 6, 24-33, 2003.

439b Hydrogel Protection Strategy to Stabilize the Water-Splitting Photoelectrodes

Jeiwan Tan

Photoelectrochemical water splitting is an attractive solar-to-hydrogen pathway. However, the lifetime of photoelectrochemical devices is hampered by severe photocorrosion of semiconductors and instability of co-catalysts. Here we report a strategy for stabilizing photoelectrochemical devices that use a polyacrylamide hydrogel as a highly permeable and transparent device-on-top protector. A hydrogel-protected Sb_2Se_3 photocathode exhibits stability over 100 h, maintaining ~70% of the initial photocurrent, and the degradation rate gradually decreases to the saturation level. The structural stability of a $Pt/TiO_2/Sb_2Se_3$ photocathode remains unchanged beyond this duration, and effective bubble escape is ensured through the micro gas tunnel formed in the hydrogel to achieve a mechanically stable protector. We demonstrate the versatility of the device-on-top hydrogel protector under a wide electrolyte pH range and by using a SnS photocathode and a $BiVO_4$ photoanode with ~500 h of lifetime.

439c UK-Hyres: The UK Hub for Research Challenges in Hydrogen and Alternative Liquid Fuels

Rachael Rothman

UK-HyRES is the flagship UK hub which aims to identify, prioritise and solve research challenges in hydrogen and alternative liquid fuels. The presentation will give an overview of the UK landscape in hydrogen research and demonstrators, then will go into detail of projects within the HyRES hub. The UK-HyRES hub is structured with four technical themes (production, storage & distribution, end use and alternative fuels) and four cross-cutting themes (economic, environmental, social and safety). Initial projects include hydrogen storage materials, particularly aimed at aviation, the electrochemical production of green ammonia and a comparison of electrolyser technologies. The first hub flexi fun call attracted 160 unique Expressions of Interest, totalling over £40m, all of which aimed to address identified priorities. This highlights the growing interest and need for research within the hydrogen and alternative liquid fuels space.

439d Integrated Co-Optimization of a Water-Power-Hydrogen System Using Renewable Energy Conversion

Minzhuo Huang, Ricardo Pinto de Lima, Omar Knio

As the energy transition evolves, hydrogen is considered an alternative energy carrier [1]. However, traditional hydrogen production processes, like the steam reforming process, involve carbon dioxide emissions. Green hydrogen is a potential alternative to address the emissions issue since it is an endogenous near zero-carbon process by using renewable energy and electrochemical water-splitting technology. However, due to the cost and efficiency of commercial electrolyzers and the uncertainty and variability of renewable energy, the price of green hydrogen is currently not competitive for worldwide usage. In this work, we propose a novel green hydrogen design and operational model integrating the operations of electrolyzer system with water desalination and renewable electricity generation. The model involves the capacity design of the proton exchange membrane (PEM) electrolyzer system, hydrogen storage, reverse osmosis desalination, water tanks, wind and photovoltaic power generation, concentrated solar power (CSP), and battery storage. In particular, the CSP plant is modeled considering details involving losses, startup, and shutdown operations. This technology set is appropriate for countries with high solar radiation and limited water availability. The model assumes an hourly resolution and one-year horizon to capture weather variability. Two types of hydrogen demand are considered: 1) local continuous demand at each hour and 2) a weekly or seasonal demand that represents, for example, international demand every period to be supplied to shipping vessels. The latter demand profile motivates the usage of hydrogen tank, desalinated water tank and thermal energy storage as buffers to take advantage of the variable renewable sources. The main objective is to minimize the levelized cost of hydrogen (LCOH) and determine the optimal capacities and operations of the technologies. Ultimately, the results can support decision-makers in designing and pursuing a sustainable and commercial multi-year plan for green hydrogen production. Case studies in the United States and Saudi Arabia indicate that the well-designed combination of all technologies can lead to an LCOH as low as 5 $/kg.

[1] Riera J.A., Lima R.M., Knio O.M., 2023. A review of hydrogen production and supply chain modeling and optimization. International Journal of Hydrogen Energy 48 (37), 13731–13755.

439e Harnessing Excess Renewable Electricity for Green Hydrogen Production: Innovative Strategies from Three Continents

Pouya Ifaei, Jonggeol Na

The quest for a net-zero future has positioned green hydrogen production at the forefront of sustainable energy solutions[1]. However, the widespread adoption of green hydrogen faces significant barriers, primarily due to its high production costs and reliance on electricity and fresh water resources[2]. This challenge is further exacerbated by the inefficiencies in the current energy systems, where substantial amounts of electricity generated from renewable sources are wasted due to the inherent mismatch between supply and demand[3]. This mismatch represents an opportunity for harnessing this renewable energy for the production of energy-intensive products like green hydrogen within a circular integration framework[4]. Acknowledging the critical role of optimizing the use of excess renewable electricity, this study introduces an innovative approach aimed at reshaping energy demand through the strategic use of media platforms. This method seeks to maximize green hydrogen production by effectively managing and utilizing surplus electricity. Our research focuses on three geographically and demographically diverse case studies: Hawaii, USA; Incheon, South Korea; and Melbourne, Australia. These locations are chosen due to their distinct electricity demand profiles, influenced by local weather patterns, demographic factors, and regional policies[5]. To explore the potential of media in influencing energy consumption behaviors, we conducted a comprehensive intercontinental field survey. The survey aimed to assess the impact of both traditional and social media on consumer energy use, particularly during peak demand periods. We collected responses from 640 participants regarding their household energy consumption following exposure to energy-saving content through popular media platforms. The data gathered enabled us to model various demand patterns, including cut-and-save, flattened-by-consumers, and flattened-by-AI strategies. These patterns illustrate different approaches to reducing and shifting energy demand, either directly by consumers or through AI-driven interventions[6,7]. By integrating these demand patterns with simulations of renewable energy production systems—comprising solar photovoltaics, wind turbines, and battery storage—we evaluated the potential for annually recovering excess electricity. This surplus electricity was then allocated to power electrolyzers for green hydrogen production[8]. Our findings reveal that media-driven demand control can significantly enhance the efficiency of green hydrogen production by optimizing electricity consumption patterns. The effectiveness of these strategies, however, varies across the studied regions, highlighting the complex dynamics between renewable energy supply and demand. By expanding this study to include additional regions, we can further refine and adapt this model, offering a scalable and efficient strategy for leveraging excess renewable electricity.

1. Dincer I, Acar C. Review and evaluation of hydrogen production methods for better sustainability. *Int J Hydrogen Energ.* Sep 14 2015;40(34):11094-11111.
2. Ishaq H, Dincer I, Crawford C. A review on hydrogen production and utilization: Challenges and opportunities. *Int J Hydrogen Energ.* Jul 22 2022;47(62):26238-26264.
3. Liu LB, He G, Wu MX, et al. Climate change impacts on planned supply-demand match in global wind and solar energy systems. *Nat Energy.* Aug 2023;8(8):870-880.
4. Lim JY, Safder U, How BS, Ifaei P, Yoo CK. Nationwide sustainable renewable energy and Power-to-X deployment planning in South Korea assisted with forecasting model. *Appl Energ.* Feb 1 2021;283.
5. Staffell I, Pfenninger S, Johnson N. A global model of hourly space heating and cooling

demand at multiple spatial scales. *Nat Energy*. Sep 14 2023.
6. Dranka GG, Ferreira P, Vaz AIF. Co-benefits between energy efficiency and demand-response on renewable-based energy systems. *Renew Sust Energ Rev*. Nov 2022;169.
7. Oliveira VRJ, Tenfen D, Fernandes RC. Demand side management in Brazil: Brief history, lessons learned, status, challenges, and trends. *Renew Sust Energ Rev*. Sep 2023;183.
8. Ba-Alawi AH, Ifaei P, Yoo C. Reliable operation of an optimal self-scheduling multicarrier system with two alternatives: Multiple renewable utilities or hydrogen facilities. *J Clean Prod*. Sep 25 2023;420.

439f Electronic Structure Optimization of Metal-Phthalocyanine Via Confining Atomic Ru for All-pH Hydrogen Evolution

Yang Hou, Zhenhui Kou

The eco-friendly production of hydrogen in actual working conditions is a promising method to address the energy crisis. Facing the gap of hydrogen evolution reaction kinetics in different conditions, recently, the introducing of noble-metal single atoms has been considered an effective way to boost electrocatalytic activity. However, electrochemical stability is poor in different conditions and the loaded metals are prone to precipitation and agglomeration on traditional substrates (such as metal oxides or hydroxides, etc.), and the complex composition of conventional supports hinders the exploration of structure-activity relationship. Meanwhile, the lack of an effective proton source and the sluggish electrolyzed water process have resulted in a large kinetics difference in hydrogen evolution at different pH conditions. Herein, we construct a two-dimensional metal-polyphthalocyanine (NiFe PPc) with clear and stable networks confining atomically Ru, through electrochemical deposition method. Benefiting from the electronic coupling effect from Ru-N bridges on NiFe PPc, the Ru_{SA}@NiFe PPc shows remarkable all-pH HER activity, featured by a low overpotential of 12 mV to reach 10 mA cm^{-2} current density, which is nearly 50 times smaller than pristine NiFe PPc and outperforming the benchmark Pt/C catalyst. We further demonstrate that a proton exchange membrane water electrolysis (PEMWE) with Ru_{SA}@NiFe PPc and commercial IrO_2 anode achieves a current density of 2.0 A cm^{-2} at a low voltage of only 2.27 V. And more importantly, we further demonstrate an alkaline electrolyzer with Ru_{SA}@NiFe PPc and stainless steel mesh (SSM), which displays excellent performance with achieving current densities of 2.0 A cm^{-2} at cell voltages of 1.72 V. Theoretical calculations reveal that the electronic structure optimization through Ru-N bridge induces the acceleration of activation/dissociation water and adsorption of $^{*}H$, thus breaking the deadlock of all-pH HER kinetics.

439g Multi-Scale Testing of Solid Oxide Electrolysis Cells, Stacks, and Systems

Jeremy Hartvigsen, Micah Casteel, Nicholas Kane, Amey Shigrekar, Joshua Gomez, Cameron Priest

The production of green hydrogen is a nascent field with many emerging technologies. T The scale of green hydrogen for existing and anticipated applications is to increase from 1 million metric tons per year to 10 million metric tons. The immediate need for the increased capacity of green hydrogen production to meet demand this demand requires substantial investment in all available technologies to decarbonize as rapidly as possible. A significant amount of progress has been made over the past several years to improve the economics and reliability of solid oxide electrolysis cell (SEOC) technology. These improvements in SOEC technology allow for the technology to meet some of the increasing demand for green hydrogen and decarbonizing the hydrogen sector. The work being done at Idaho National Laboratory is an effort to validate and provide reliable third-party data on the performance and endurance of current commercial stack offerings. Stack testing at scales between 3 kW and module testing at levels of 100 kW are discussed. Button cell operational data and degradation rates up to 3,000 hours are compared favorably with stack operational data up to 4,000 hours. Through cost sensitivity analysis, we highlight the most significant challenges observed in hydrogen cost and discuss the impact of stack longevity at full commercialization. Finally, we demonstrate stack operation under dynamic load conditions to better reflect the expected operational characteristics needed to match green electricity. At the stack level, a commercial supplied has demonstrated multiple H2 production cycles ramping from 90% load to no load in less than 10 minutes. The stack also operated with instantaneous load and unload cycles. Through the comparisons of operational scales, we conclude that dynamic operation of cells, stacks and systems is feasible. The dynamic operation provides economic advantage to assist in meeting technical targets set by the Department of Energy in pursuit of the hydrogen Earthshot goal of $1/kg by 2031.

439h Optimizing Renewable Hydrogen Production: Insights from Regional Variabilities and Storage Solutions in the U.S Regions.

Bosong Lin, Paul Sizaire, Emre Gencer

In the face of a global surge in hydrogen demand, the strategic design and optimization of the hydrogen supply chain become crucial. This study focuses on the renewable hydrogen supply chain across key U.S. regions, including Gulf Coast, California, and Midwest. Our analysis evaluates the Levelized Cost of Hydrogen (LCOH) across the supply chain, shedding light on the impact of techno-economic assumptions for renewable energy sources and electrolysis, the strategic use of geological salt caverns for storage, and the diversity in renewable energy generation mixes.

Our findings reveal a dominant production cost, which accounts for nearly 90% of the total LCOH, with initial costs in 2025 projected at \$6.1/kg, \$7.8/kg, and \$6.3/kg for the Gulf Coast, California, and Midwest, respectively. The study anticipates a potential reduction in LCOH by a minimum of 52% by 2050, thanks to technological advancements and efficiency gains in renewable energy and electrolysis processes. A detailed comparison between salt cavern storage and liquid hydrogen storage models highlights a possible increase in LCOH by approximately \$1.0/kg for scenarios dependent on liquid storage.

Moreover, our investigation into the system cost implications of renewable energy mixes emphasizes the critical importance of oversize factors. A one-to-one ratio of renewable energy to electrolysis capacity could elevate the LCOH to \$11.3/kg, \$11.5/kg, and \$14.7/kg for the respective regions under a pure solar PV or onshore wind framework. These insights underline the significance of oversizing factors and diverse energy mixes in crafting efficient renewable hydrogen systems, paving the way for informed industrial applications. Additionally, the exploration of offshore wind as a component of the hydrogen production mix underscores its current economic challenges. Without substantial cost reductions, offshore wind remains an economically unfeasible option. Our study contributes to a deeper understanding of the factors influencing the cost-efficiency and sustainability of renewable hydrogen production, offering valuable perspectives for future development and deployment strategies.

AUTHOR INDEX